Onno Bakker   **Lumi-Imager™ F1 –** *Lab Protocols*

Springer-Verlag Berlin Heidelberg GmbH

Onno Bakker

Lumi-Imager™ F1
Lab Protocols

With 14 Figures,
partly in Color

 Springer

ONNO BAKKER, Ph.D.
Academic Medical Centre
Endocrinology, F5-171
Meibergdreef 9
1105 AZ Amsterdam
Netherlands

ISBN 978-3-540-64794-2   ISBN 978-3-662-08431-1 (eBook)
DOI 10.1007/978-3-662-08431-1

Library of Congress Cataloging-in-Publication Data applied for.

Die Deutsche Bibliothek – CIP-Einheitsaufnahme
Bakker, Onno: Lumi-imager F1 – lab protocols/Onno Bakker (author). – Berlin; Heidelberg, New York; Barcelona; Budapest; Hong Kong; London; Mailand; Paris; Singapore; Tokyo: Springer, 1998

© Springer-Verlag Berlin Heidelberg 1998
Originally published by Springer-Verlag Berlin Heidelberg New York in 1998.

Data conversion: K+V Fotosatz, Beerfelden
Cover Design: Mitterweger Medien GmbH, Plankstadt

SPIN 10687024   18/3137-5 4 3 2 1 0 – Printed on acid-free paper

Trademarks

- Lumi-Imager™, LumiAnalyst™, TriPure™ are trademarks of Boehringer Mannheim GmbH
- Windows™ and Microsoft® Excel are a trademark and registered trademark of Microsoft Corp.
- CSPD®, CDP-*Star*™, Western-Light™ and Nitro-Block™ are a registered trademark and trademarks of Tropix Inc.
- Stratalinker® and PosiBlot® are registered trademarks of Stratagene Inc.
- Saran Wrap® is a registered trademark of Dow Chemicals
- ECL™ is a trademark of Amersham International Plc
- SYBR® green and SYPRO® red are registered trademarks of Molecular Probes
- Protran® BA85 is a registered trademark of Schleicher and Schuell
- PolyATract® is a registered trademark of Promega Corp.

1 Digital imaging and quantification of chemiluminescence and fluorescence: Lumi-Imager and its applications*

Onno Bakker[1], Marti Aldea[2], Jörg Bergemann[3],
Albert Geiger[4], Michael Kirchgesser[5], Achim Kramer[6],
Stephan Schröder-Köhne[7], Hans-Joachim Höltke[8]

ABSTRACT

The instrument, Lumi-Imager was developed for the detection and accurate quantification of chemiluminescent signals on blots and in microtiter plates. It contains a cooled CCD camera, a specially developed lens system and software. In addition to luminescence, Lumi-Imager can also detect and quantify UV-excited fluorescent signals in gels. Here we describe the components and features of this instrument, as well as some of its applications.

* Parts of this article have been published in Onno Bakker et al., "Digital imaging and quantification of chemiluminescence and fluorescence: detection of mutation with protein truncation test". In: Practical use of non-radioisotopic systems. Part II – commercially available kits, ed. S. Nomura and T. Watanabe (Tokyo: Shujunsha, 1998), 140–154

[1] Laboratory of Experimental Endocrinology, F5–171, Academic Medical Centre, Meibergdreef 9, 1105 AZ Amsterdam, The Netherlands
[2] Dept. Ciències Mèdiques Bàsiques, Universitat de Lleida, Rovira Roure, 44, 25198 Lleida, Catalunya, Spain
[3] Paul Gerson Unna Skin Research Center, Beiersdorf AG, Unnastraße 48, 20245 Hamburg, Germany
[4] PNA Diagnostics A/S, Rønnegade 2, DK-2100 Copenhagen Ø, Denmark
[5] Boehringer Mannheim GmbH, Nonnenwald 2, 82372 Penzberg, Germany
[6] Institut für Medizinische Immunologie, Universitätsklinikum Charité, Schumannstraße 20–21, 10098 Berlin, Germany
[7] Max-Planck Institut für biophysikalische Chemie, 37070 Göttingen, Germany
[8] Boehringer Mannheim GmbH, Nonnenwald 2, 82372 Penzberg, Germany

AIM

The aim of the authors is to present the reader with their experience in using Lumi-Imager, an apparatus recently developed by Boehringer Mannheim to accurately quantify „glow-type" non-radioactive luminescent signals.

INTRODUCTION

Non-radioactive methods are increasingly used in many laboratories and are starting to replace traditional methods which involve radioactive isotopes for the labeling and detection of nucleic acids and proteins. Non-radioactive methods not only avoid the hazards, cost, and inconvenience associated with radioactive methods, but are also often faster and more sensitive. Enzyme-catalyzed chemiluminescence has established itself as the most sensitive non-radioactive detection method. Initially, chemiluminescence was widely used in Western blotting applications. Non-radioactive haptens, such as digoxigenin (DIG) or vitamin H (biotin), were later established for nucleic acid labeling, whereby chemiluminescence proved to be highly sensitive in applications such as Southern, Northern, dot blotting and gel-shift assays. A disadvantage of these non-radioactive chemiluminescent methods was that the signals had to be detected with X-ray films. Due to their very low dynamic ranges (approximately 1:100), these films cannot simultaneously image very strong and very weak signals without overexposure of the strong signal. Additionally, detecting signals with X-ray films requires the use of a dark room, exposure to be set up in a light-tight cassette, and films to be later developed. This procedure entails extra work and inconvenience, and also involves the handling of hazardous chemicals. Furthermore, it is not possible to directly quantitate signals on X-ray film. Indirect quantitation is achieved by scanning the films, after which absorbance values can only then be taken as an indirect estimation of the intensity of luminescence. As X-ray films exhibit a non-linear response to luminescence (as well as to radioactivity), data acquired in this manner can only provide a rough approximation of the true light emission from the object. Nowadays, CCD cameras offer the possibility of sensitive detection and direct quantitation of light. In fact, deeply cooled CCD cameras are used in astronomy to observe faint, distant objects.

Based on this technology, Boehringer Mannheim has developed Lumi-Imager, a cooled CCD camera and special lens system for the highly sensitive detection and accurate quantitation of chemiluminescence on blots and in microtiter plates.

1.1 Technical description of the instrument

The system consists of a Peltier-cooled CCD camera (cooled to about –40 °C) and a specially developed lens system that allows image acquisition of „glow-type" chemiluminescent signals, e.g., catalyzed by alkaline phosphatase or horseradish peroxidase on samples of up to 24×30 cm^2 or four microtiter plates (Fig. 1). The drawer can accommodates an UV-transilluminator, in which case a filter wheel with four different bandpass filters is mounted in front of the lens system. The high resolution detector (1280×1024 pixels) and specially designed optics achieve a resolution of two lines per millimeter over the whole

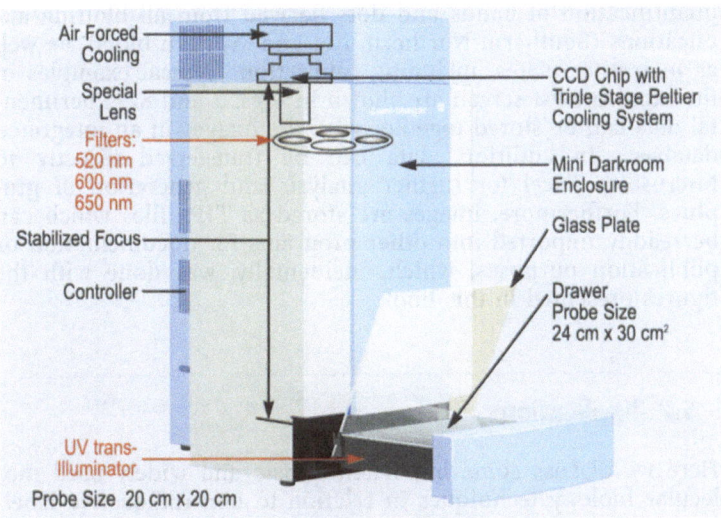

Fig. 1. Photograph and schematic drawing of the Lumi-Imager instrument

sample area (24×30 cm²) and permit a fixed optical alignment. This fixed optical alignment, together with a specially designed focus stabilization which compensates for temperature variations and long-term exposure effects, avoids tedious and repeated focusing. These special optics and the built-in Windows-based camera driver software, which automatically performs flatfield- and detector-specific corrections, enable accurate quantitation of chemiluminescent signals across the entire sample area. For microtiter plate measurements, Lumi-Imager demonstrates excellent linearity over the whole dynamic range and compares very favorably to dedicated luminescence microtiter plate readers. With regards to contrast and resolution, sensitivity and image quality are wholly comparable to X-ray film.

In addition, Lumi-Imager provides the option to use a more sensitive „binning" mode which reduces exposure time by a factor of up to four compared to X-ray film; and although resolution is reduced two-fold in the process, it is still comparable to that of most phosphoimagers. The instrument has a dynamic range of 1:10 000 which allows for imaging and quantitative analysis of weak and strong signals in a single exposure. To this end, LumiAnalyst software has been developed which enables fast and accurate molecular weight determination and quantification of bands and dots derived from all blotting applications (Southern, Northern, dot and Western blots), as well as microtiter plates, including calibration. Typical examples of the LumiAnalyst screen are shown in Figs. 7 and 8. Experimental data can be stored together with the images in an integrated database. In addition, data can be transferred directly to Microsoft® Excel for further analysis and generation of graphics. Furthermore, images are stored as TIFF files which can be readily imported into other programs for documentation or publication purposes, which, incidentally, was done with the figures presented in this book.

1.2 Applications

Here we discuss some important classic and widely used molecular biology techniques in relation to non-radioactive labeling and signal detection/quantification using Lumi-Imager. In addition, more recent and exciting methods, such as the pro-

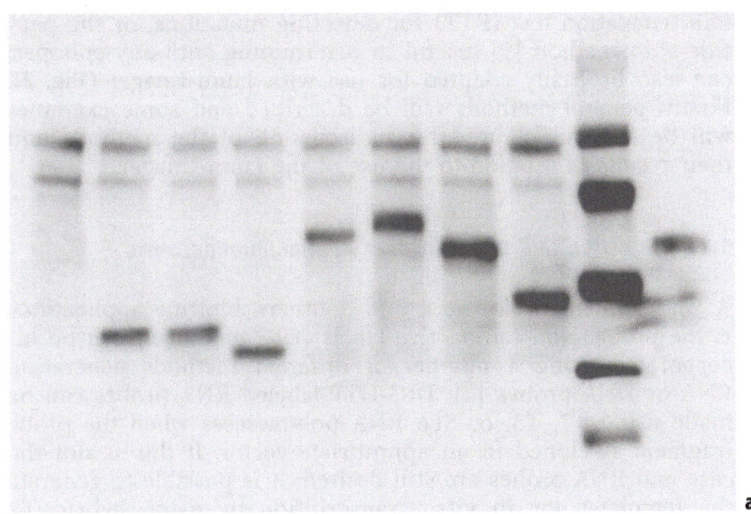

a

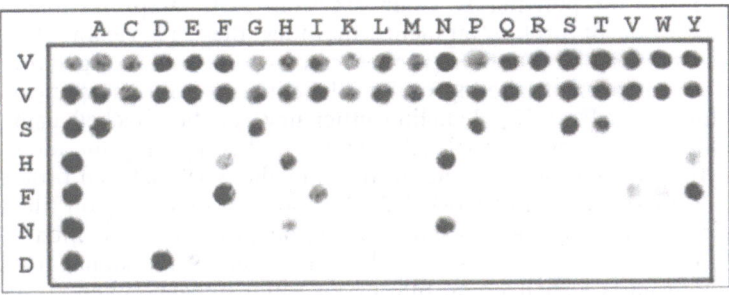

b

Fig. 2 a, b. Examples of emerging techniques suited to the Lumi-Imager. **a** An example of a typical protein truncation test. In this case, fragments of the adenomatous polyposis coli coding region of familial adenomatous polyposis patients were amplified by polymerase chain reaction from genomic DNA, thereby incorporating a T7 promoter at the 5′-end. After in vitro transcription and translation (using the Protein Truncation Test Kit, non-radioactive, Boehringer Mannheim) the resulting protein was analyzed by gel electrophoresis. Truncated translation products indicate the presence of a stop mutation. The focus was on exon 15, known to contain the majority of mutations in a so-called mutation cluster region. **b** An example of a peptide scan; in this case, substitutional analysis of the TGFα epitope, VVSHFND, as recognized by the monoclonal antibody, Tab. 2., of a substitutional analysis of the peptide VVSHFND. Each position of the peptide is substituted by all 20 amino acids (*rows*). *Spots* in the *left column*, acquired with Lumi-Imager, are identical and represent the wild-type peptide. This experiment shows that C-terminal asparagine (*N*) and aspartic acid (*D*) are essential for antibody recognition. In contrast, the two N-terminal valine (*V*) residues can be replaced by all other amino acids without loss of antibody binding

tein truncation test (PTT) for detecting mutations, or the peptide scan method [9], useful in determining antibody epitopes, can also be easily adapted for use with Lumi-Imager (Fig. 2). Firstly, general methods will be described and some examples will be shown, followed by a discussion of the methods and their results in relation to the use of the Lumi-Imager.

1.2.1 Detection and quantification of chemiluminescence

A key element in Northern and Southern blotting applications is the probe. Non-radioactive labels (DIG or biotin) can be incorporated using a number of different methods generating RNA or DNA probes [2]. DIG-UTP-labeled RNA probes can be made using T7, T3, or SP6 RNA polymerases when the probe fragment is cloned in an appropriate vector. If this is not the case and RNA probes are still desired, it is possible to generate the templates for in vitro transcription by using hybrid T7 primers for polymerase chain reaction (PCR) amplification of cDNA or of probe fragments in other than T7 vectors (see Chapter 3 and 4). DNA probes can be prepared using either random primed labeling or PCR labeling, in both cases incorporating DIG-dUTP. Labeling efficiency can be checked using dilutions of known labeled fragments and comparing the intensities of the different spots to those of the newly labeled probe. Alternatively, 1 µl of labeled RNA is spotted onto a nylon filter, fixed and, using a shortened detection procedure (all incubations of 10 min or less, see Chapter 3 and 4), the signal from this spot is visualized on film or via Lumi-Imager. When an intense black spot is visible after an 1-min exposure, the probe is considered suitable for further use.

Filters can be prehybridized and hybridized in a number of different solutions [2]. One of these is DIG Easy Hyb, an urea-based hybridization solution. Probes are at a concentration of between 20 and 100 ng/ml hybridization mix. When using a new probe or batch of membrane, it is wise to perform a „mock" hybridization to determine the maximum amount of probe that can be used [2]. An important advantage of non-radioactive probes is that the hybridization mix containing the probe can be frozen at −20 °C and reused at least four times. Hybridization temperature is between 42 °C and 65 °C, depending on the probe used. When using RNA probes, it is advisable to have a stringency of two washes in 0.1×sodium saline citrate (SSC)/0.1% sodium dodecyl sulfate (SDS) at 65 °C (Northern) or 50 °C (Southern). This may also work for DNA probes, but sometimes 0.5×SSC/0.1% SDS gives better results at the hybridization temperature. Detection is performed using CSPD or CDP-Star) as the substrates for alkaline phosphatase [2]. The final image can then be recorded with Lumi-Imager and all the advantages it has to offer, as will be pointed out below. Typical exposure times are between 5 min and a few hours when using total RNA or genomic DNA, but can be less then 5 min when using polyA$^+$ RNA.

1.2.2 Northern blotting

Northern blotting is an important technique for studying gene expression in different tissues and species and signal quantification is usually of great importance. It is a technique well suited to non-radioactive labeling techniques [3–5, 10, 11, 15]. Many methods have been described for Northern blotting [1]. A typical method is provided below (Chapter 4) which has been successfully used in studying gene expression in mammalians and yeast. Examples of Northern blots generated from these species using this method are presented in Figs. 3–5.

Total RNA and/or polyA$^+$ RNA can be prepared from tissue or cells using different reagents or methods [1]. Samples are preferably run on formaldehyde-agarose gels and approximately 1–2 µg polyA$^+$ or 1–10 µg of total RNA per well is usually applied. After completion of the run, the gel can be blotted on nylon membrane (i.e., Boehringer Mannheim or another supplier) using a number of different methods, taking care that formaldehyde is removed by incubating the gel for 10–20 min in 20×SSC before blotting. RNA can be fixed on the blot by baking it for 30 min at 120 °C or by UV irradiation, preferably using specialized equipment (i.e., Stratalinker, Stratagene). Other features of these methods have been described above.

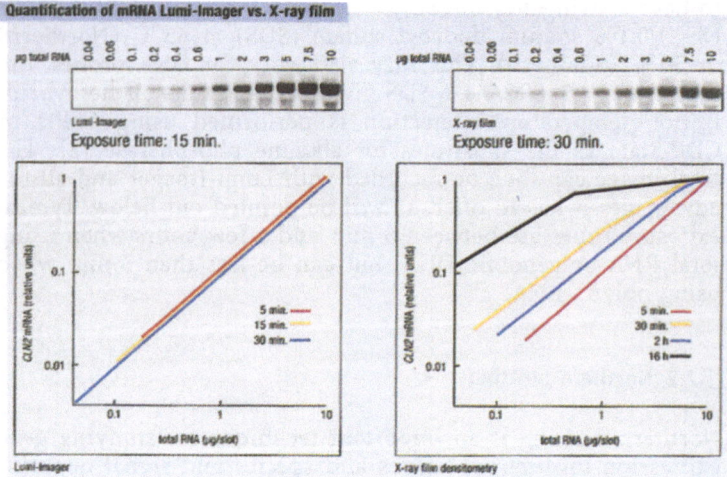

Fig. 3. Quantification of yeast mRNA using Lumi-Imager vs. film. Different amounts of yeast total RNA where used to quantify mRNA levels of the G_1-cyclin *CLN2* gene (not a highly expressed gene in yeast, *Saccaromyces cerevisae*) with a DIG-labeled probe. The figure shows the relation between the intensities measured at different exposure times in relation to the total amount of RNA applied (*left panel*), and a comparison of the results obtained by direct capture with the Lumi-Imager and by scanning X-ray films exposed for different lengths of time (*right panel*)

1.2.3 Southern blotting

Southern blotting is another important molecular biology technique which can also be readily performed with non-radioactive labels. Similar to Northern blotting, many methods have also been described for Southern blotting [1]. A typical method which has been successful in studing human gene is described in Chapter 3 (see Chapter 3, Fig. 2 for an example).

DNA (genomic or other) can be prepared from tissue or cells using a variety of methods [1]. Samples are run on agarose gels and for studying mammalian genes, 5–10 µg genomic DNA per well is usually applied. After completing the run, the gel can be blotted on nylon membrane (i.e. Boehringer Mannheim or another supplier) using a number of different methods, taking care that, where necessary, gels are treated with 0.1 N HCl and neutralized [1]. To fix DNA, similar methods to those described for Northern blotting can be used.

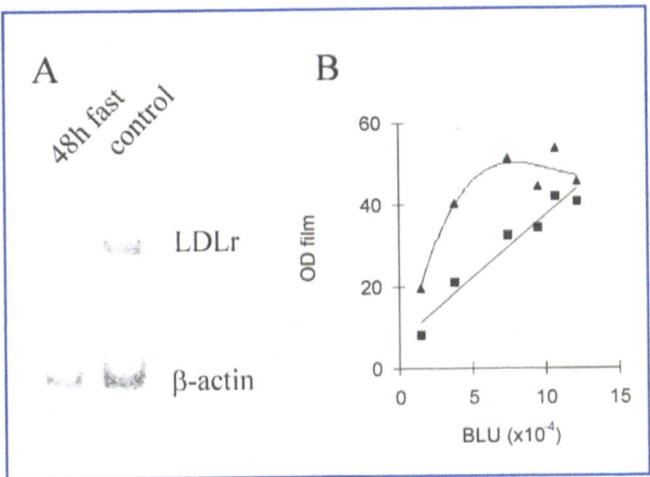

Fig. 4a, b. Quantification of rat mRNA using Lumi-Imager vs. film. **a** Low-density lipoprotein (*LDL*) receptor (45-s exposure) and β-actin (5-s exposure) mRNA expression in livers of control rats or fasted (48 h) rats (polyA$^+$ RNA). **b** Curves depicting the relationship between the LDLr mRNA density of a 45-s (*black squares*) or 2-min (*black triangles*) exposure on film and direct measurement of chemiluminescent signals by Lumi-Imager. Curve fitting parameters: 45 s, $y = 3.04 \times + 7.2$, $R^2 = 0.94$; 2 min, $y = 0.05 \times^3 - 1.60 \times^2 + 15.70 \times + 0.47$, $R^2 = 0.93$ (a linear fit gave a R^2 of 0.44)

Special Southern-like assays based on test strips and microfabricated devices containing arrays of immobilized nucleic acids have recently been developed for a variety of uses in genomic analysis [6, 8]. One of the more recent additions to this type of assay involves peptide nucleic acid (PNA), a DNA mimic that shows some unique properties compared to DNA itself. Due to its neutral backbone, PNA can hybridize to nucleic acids in the absence of counterions which are normally needed to stabilize hybrids (in „normal" Southern blotting) between DNA probes and single-stranded DNA or RNA targets [7, 13]. In addition to this property, PNA exhibits better hybridization characteristics (affinity and specificity) compared to DNA [7, 12]. Like DNA probes, PNA probes can be used in both homogeneous and heterogeneous assay formats.

Regular arrays of immobilized nucleic acids or PNAs are ideally suited to the Lumi-Imager as a hybridization detection tool (Fig. 7). Due to its sensitivity and the LumiAnalyst soft-

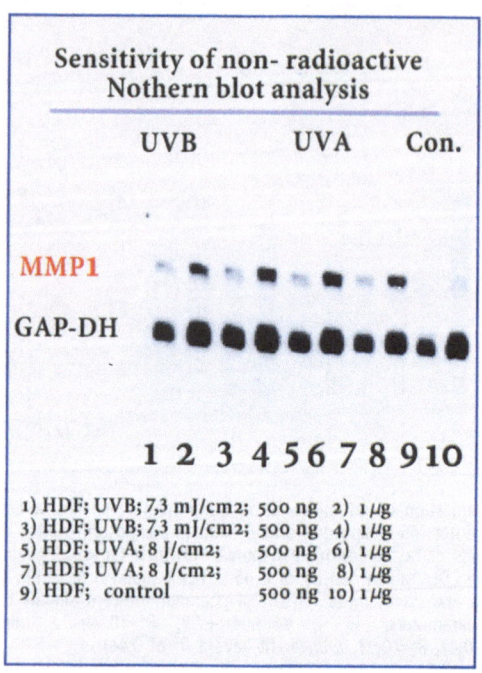

Fig. 5. Expression of matrix metalloproteinases (*MMP*) in human skin biopsies. There is clear correlation between concentrations of UVA filters (from *left* to *right*) and protection against MMP1 induction. Expression of MMP1 mRNA (*top band*) can be measured in as little as 500 ng RNA (30 min exposure on the Lumi-Imager). The *bottom band* is an internal control with GAP-DH

ware, rapid (automated) analysis of the resulting hybridization patterns is possible.

1.2.4 Western blotting

Western blotting generally involves the application of protein samples to an SDS-PAGE gel of a strength usually between 6% and 12%. Once separation is completed, the gel must be transferred to a membrane which can be either nitrocellulose or PVDF. For very accurate quantification, PVDF membranes are recommended. Transfer is usually performed using either „wet"

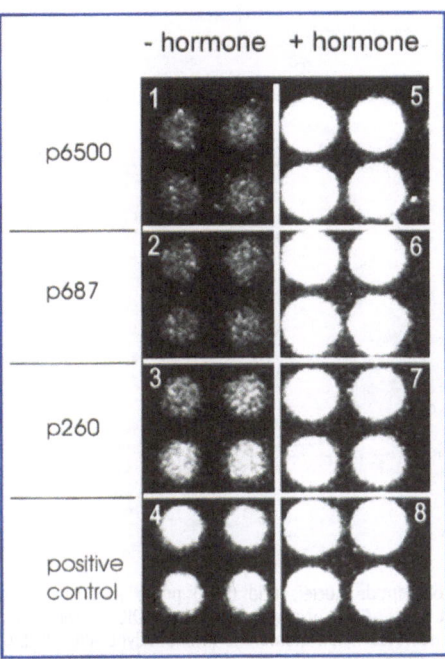

Fig. 6. Thyroid hormone responsiveness of the human LDLreceptor promoter. Hela cells were cotransfected in duplicate with a thyroid hormone receptor expression plasmid and 5′-deletions of the human LDL-receptor gene promoter (to –6500, –687 and –260 bp respectively) cloned in front of secreted alkaline phosphatase (SEAP). After transfection cells were incubated with (+ hormone) or without (– hormone) thyroid hormone and after 48 h SEAP activity was measured in duplicate using the SEAP reproter gene assay kit (Boehringer Mannheim). Detection and quantification were done using Lumi-Imager (5 min exposure; for more data see [16]).

or „semi-dry" electroblotting. After transfer is completed and has been verified, using for instance Ponceau S staining, the blot is blocked and then incubated with the first antibody. Here, little detail is provided regarding blocking solutions and dilutions of primary and secondary antibodies, the success of which depends to a large extent on the particular protein and antibody in question (see Chapter 2). It is suggested that the reader studies a number of methods to find the one suited to his/her particular case [1, 9]. Signals can be detected using either alkaline phosphatase- or horseradish peroxidase-labeled

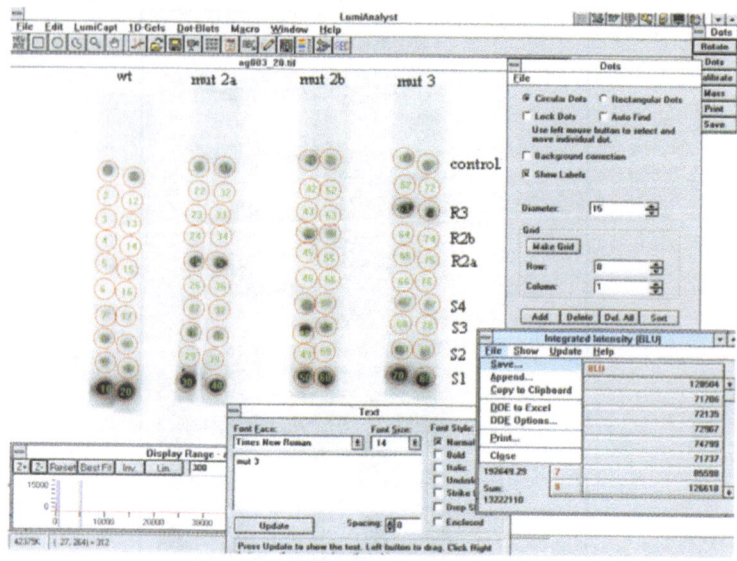

Fig. 7. Analysis of peptide nucleic acid (PNA) probe specificity; a typical LumiAnalyst screen in dot-blot mode. A set of seven PNAs and a DIG-labeled molecule (*control*) were pipetted onto an activated nylon membrane (Immunodyne ABC; PALL Corp.) and immobilized covalently by UVirradiation. The set of PNAs consisted of four probes (*S1–S4*), specific for the wild-type form of the target, and three probes specific for relevant mutations in the corresponding regions (*R2a, R2b*, and *R3*). The membrane was subsequently challenged with either DIG-labeled wild-type oligonucleotides (*wt*) or oligonucleotides containing single base mutations (*mut2a, mut2b, mut3*). After the subsequent DIG detection process [2], chemiluminescent signals were analyzed using the LumiAnalyst software. Readouts for the seven probes were used to calculate the cross-hybridization rate for single base mutations, and signals from the control spot acted as a built-in control for the spotting process

secondary antibodies, and the CSPD/CDP-Star substrates or luminol-type substrates, respectively (for an example, see Fig. 8).

1.2.5 Microtiter plate assays

Lumi-Imager is well suited to the detection and quantification of chemiluminescent signals in microtiter plates [16]. A number of assays for β-galactosidase, chloramphenicol acetyl transferase (CAT), and secreted alkaline phosphatase (SEAP), which

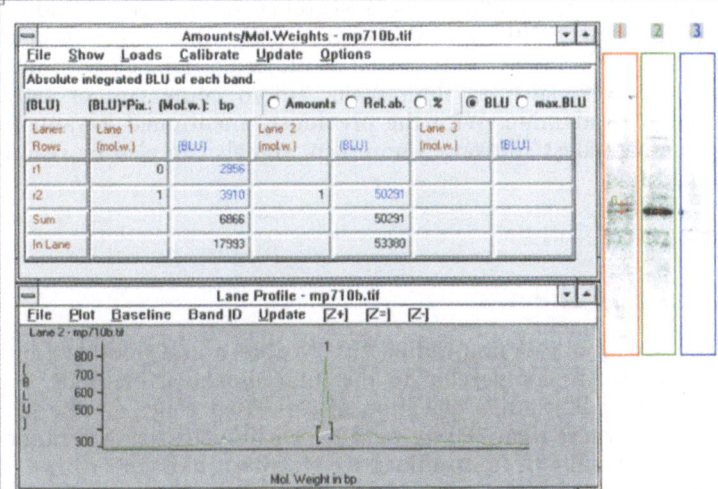

Fig. 8. Western blot of rat liver nuclear extracts; a typical LumiAnalyst screen in 1-Dgel mode. Nuclear extracts from rat liver (25 µg) were run on a 10% SDS-PAGE gel and the gel was blotted onto nitrocellulose (BA45, Schleicher and Schuell). First antibodies were specific for the $\alpha 2$, $\beta 1$, and $\beta 2$ thyroid hormone receptor isoforms (lanes *1*, *2*, and *3*, respectively). Secondary antibody was alkaline phosphatase conjugated to goat-antirabbit IgG. Detection was with CDP-Star and exposure time on Lumi-Imager was set at 30 min. TR$\alpha 2$ is expressed about ten-fold lower than $\beta 1$ (in agreement with mRNA data), whereas $\beta 2$ is not expressed in rat liver

are routinely used in promoter/expression studies, have been or can be adapted for use with Lumi-Imager (Fig. 6). Basically, any microtiter plate-based assay that can generate „glow-type" chemiluminescence as its readout is adaptable to the apparatus. With LumiAnalyst's programmed-in 8×12 well mask a readout can be obtained with Lumi-Imager within minutes. This data can then be exported to Excel, yielding final results within another few minutes.

1.2.6 Detection and quantification of fluorescence

In addition to luminescence, Lumi-Imager is also able to detect and quantify UV-excited fluorescent signals in gels, when an UVtransilluminator is placed in the drawer of the apparatus.

This new option allows the researcher to quantify Ethidium bromide or SYBR green-stained nucleic acids, or proteins stained with SYPRO red, as well as molecules labeled with fluorescent tags, such as fluorescein, amino coumarin, or tetramethyl rhodamine. With the previously mentioned advantages of LumiAnalyst software, fluorescent signals can also be readily quantified.

1.3 Discussion

The sensitivity of non-radioactive Northern and Southern blot techniques is comparable to that of radioactive methods. For example, quantitative detection of expressed genes can be performed in less than 500 ng total RNA using DIG-labeled probes (typical results of non-radioactive Northern blots are shown in Figs. 3–5). Furthermore, the sensitivity of both Northern and Southern blotting can be enhanced by using RNA probes. An additional advantage of non-radioactive labeling is the long shelf-life of DIG-labeled RNA or the DNA probes themselves (2 or more years).

One of the problems encountered when using X-ray film is that there is only a small range of signal intensities on any one blot which can be accurately quantified simultaneously. This requires that several exposures be taken and all have to be individually analyzed with regard to linearity of signal intensity – a process which is both costly and time consuming. There is good correlation between the densities measured on the „right" exposure of film and data obtained using Lumi-Imager (see Fig. 3, right panel). Furthermore, exposure times with Lumi-Imager remain similar. In the case of β-actin in rat liver, comparison of a 5-s exposure on film (the „right" one) and the same exposure with the Lumi-Imager, yields a linear relationship (y=3.3×–10.3, R^2=0.98). As can be seen in Fig. 4b, good linear relationship is obtained when the intensities of the LDLr mRNA obtained using the Lumi-Imager are compared to those obtained from scanning a 45-s exposure on film. This is evidently not the case for a 2-min exposure on film where the film is already saturated (see Fig. 4b). On the other hand, when a 5-s exposure of the β-actin signal is plotted against a 5-min exposure, both obtained using Lumi-Imager, a good linear relationship is obtained (y=10.3×–20332, R^2=0.98), indicating that there is more flexibility in exposure time when using the Lumi-Imager than when using film (see also Fig. 3, right panel).

Lumi-Imager can also be used to quantify signals on Western blots where signal is detected using luminol-type substrates. Figure 9a demonstrates that, in eight independent blots, signal linearity was very good in the range tested (5–20 µl yeast extract). From the graphs it is apparent that linearity could have been somewhat compromised at much lower loadings. Occasional deviations from the linear relationship may result from pipetting errors during the application of the sample on the SDS-PAGE gel, or may be due to localized substrate depletion (see below). Figure 9b demonstrates the linearity of signal intensity vs. exposure time. Three different blots (Fig. 9b shows a typical example) were first exposed in Lumi-Imager for 2 min and then for 10 min. Thereafter, images were analyzed as described in the legend to Fig 9a. Duplicate samples were applied in adjacent lanes of the gel. Figure 9b demonstrates that linearity is excellent, where a five-fold increase in exposure time results in five-fold higher signal intensity (compare y-axis). However, some signals occasionally show a less than linear increase with time, as is the case in Fig 9c where none of the 10-µl values increased linearly with time. This is most probably due to local differences in the concentration of luminol substrate. During exposure, the blot is sandwiched between two layers of Saran-Wrap. In this setup, air bubbles may become trapped between the blot and the plastic foil, thereby impeding uniform coverage of nitrocellulose with substrate. If air bubbles are observed, one should critically evaluate results obtained from exposure times longer than a few minutes. In order to distinguish between pipetting errors from localized substrate depletion, one should always quantify signals at two different exposure times.

From the above results, we conclude that it is possible to reliably quantify intensities of bands on Northern and Western blots using the Lumi-Imager, with the added, important, advantage that there is more flexibility in exposure time. Moreover, time spent on analyzing raw data is greatly reduced. The advantages in using Lumi-Imager are realized not only with Northern and Western blots, as detailed above, but also with other chemiluminescent assays routinely performed in the lab-

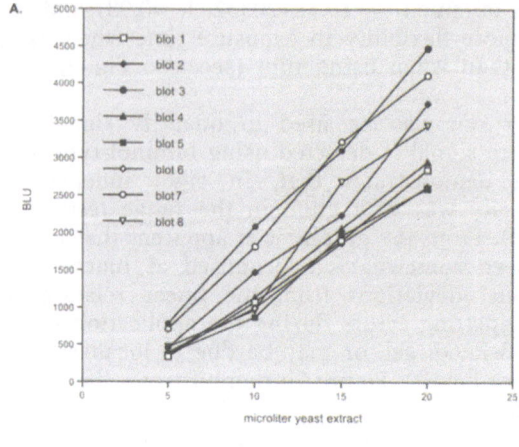

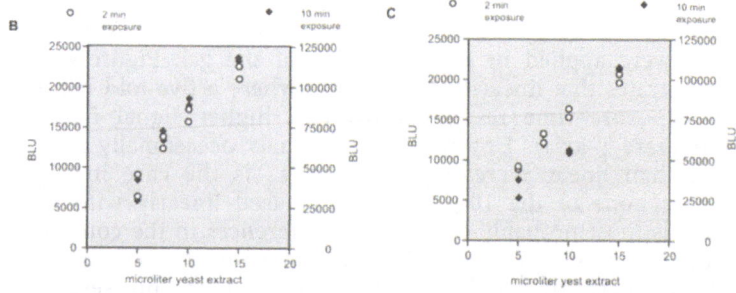

Fig. 9 a–c. Expression of the yeast protein, Emp47p, in total yeast extract; Western blot. **a** The signal derived from Emp47p was quantified using an Emp47p-specific polyclonal rabbit-antiserum. After final incubation in luminol-type substrate (1 min), the signal of each lane was immediately captured by Lumi-Imager. It was then quantified by marking the bands with a circumferential rectangle in the dot-blot mode and an identical rectangle adjacent to each signal was used to define the lane-specific background (this method could be better than alternative methods of quantification and background-correction in certain cases). **b** This data demonstrates linearity of signal intensity vs. exposure time. Blots were first exposed in Lumi-Imager for 2 min and then for 10 min. The images were then analyzed as described above. Samples were applied in duplicate in adjacent lanes of the gel. The graphs show light units (*BLU*) measured as a function of the amount of yeast extract used (*x-axis*) at the two different exposures (*left y-axis*, 2-min exposure; *right y-axis*, 10-min exposure)

oratory, such as Southern blotting, reporter gene assays, and gel-shifts.

Acknowledgments. We thank O. Strobel for critical reading of the manuscript.

References

1. AUSUBEL FM, et al. (1991) Current protocols in molecular biology. Wiley, New York
2. Boehringer Mannheim (1995) The DIG system user's guide for filter hybridisation
3. BOELEN A, et al. (1995) J Endocrinol 146:475–483
4. BOELEN A, et al. (1996) Endocrinology 137:5250–5254
5. BOELEN A, et al. (1997) J Endocrinol 153:115–122
6. EGGERS M, EHRLICH D (1995) Hematol Pathol 9:1–15, and references therein
7. EGHOLM M, et al. (1993) Nature 365:566–568
8. GOFFEAU A (1997) Nature 385:202–203, and references therein
9. HARLOW E, LANE D (1988) Antibodies, Cold Spring Harbor
10. HUDIG F, et al. (1994) FEBS Lett 341:86–90
11. HUDIG F, et al. (1997) J Endocrinol 152:413–421
12. NIELSEN PE, et al. (1991) Science 254:1497–1500
13. ØRUM H, et al. (1995) BioTechniques 19:472–478
14. SCHNEIDER-MERGENER J, et al. (1996) In: CORTESE R (ed) Combinatorial libraries. W. de. Gruyter, Berlin, pp. 53–68.
15. VAN DER WAL AMG, et al. (1998) Int J Biochem 30:209–215
16. BAKKER O, et al. (1998) Biochem Biophys Res Comm (in press)

2 Western blotting

O. Bakker

INTRODUCTION

To study protein expression, qualitatively or quantitatively, Western blotting is usually the method of choice. It allows for the detection of a protein of interest with a specific antibody on a solid support (mostly nitrocellulose or PVDF membrane). Detection of the signal can be done using radioactivity, color substrates, or the more recently developed chemiluminescent substrates. This protocol aims to give the potential user a guideline for setting up a Western blotting experiment using chemiluminescent substrates which are either based on luminol/horseradish peroxidase (HRP/POD) or dioxetane/alkaline phosphatase (AP). When used in combination with a Lumi-Imager, both these detection methods allow quick and accurate detection of the protein.

2.1 Materials

2.1.1 Equipment

- Electrophoresis equipment
- Electroblot equipment (tank or semi-dry)
- Shaking platform and rotating oven
- If available, a Lumi-Imager.

2.1.2 Materials

- Membrane; either nitrocellulose (e.g., Protran® BA85; Schleicher and Schuell) or PVDF (e.g., Boehringer Mannheim)
- Specific (first) antibody (polyclonal or monoclonal) raised against the protein of interest
- Second antibody (which recognizes the constant part of the first antibody) conjugated with HRP/POD or AP (supplied in different kits or available separately from, for instance, Boehringer Mannheim; DAKO; or Tropix)
- Chemiluminescent substrate; either luminol-type or dioxetane-type are available from various suppliers, such as for example, luminol-type substrate, ECL (Amersham International); Lumi-Light^{plus} Western blotting kit (Boehringer Mannheim, Mannheim, Germany); dioxetane-type substrate, Western-Light kit (Tropix).

2.1.3 Buffers

4×TRIS/SDS buffer 0.5 M TRIS HCl 0.4% SDS	pH 6.8

2×Loading buffer 25 ml 4×TRIS/SDS buffer 20 ml Glycerol 4 g SDS 3.1 g DTT 1 mg Bromophenol blue Fill up to 100 ml with water Store in aliquots at −20 °C	pH 6.8

1×TBS 50 mM Tris HCl 150 mM NaCl	pH 7.5

Ponceau S
Dissolve 0.5 g Ponceau S in 0.5 ml glacial acetic acid
Water to 100 ml
Best prepared fresh

1% Blocking solution (other blocking buffers can be used [1, 2])
Dissolve 2 g blocking reagent (Boehringer Mannheim) in 200 ml of
1×TBST by stirring until dissolved (about 1 h)
Stable for 1 week at 4 °C

5% Blocking solution (other blocking buffers can be used [1, 2])
Dissolve 5 g blocking reagent (Boehringer Mannheim) in 100 ml of
1×TBST by stirring until dissolved (about 1 h)
Stable for 1 week at 4 °C

1×TBST
0.1% Tween 20 in TBS

Transfer buffer
48 mM Tris base
39 mM Glycine
0.01%–0.1% SDS (see note)
15%–20% Methanol (see note)
Note: The concentration of SDS and methanol in the transfer buffer can
influence the transfer of the proteins [1]. For PVDF, use a maximum of
0.01% SDS and 15% methanol.

Assay buffer (dioxetane protocol)
100 mM Tris   pH 9.5
100 mM NaCl

2.2 Procedure

2.2.1 Protein

The source of the protein of interest can be variable (tissue, cells, blood, etc.), as are the ways of purifying it. Therefore, no general method can be given and the reader is referred to standard protocol books [1, 2] and/or specific literature on her/his protein of interest.

2.2.2 Gel electrophoresis and blotting

Although it is outside the scope of this protocol to describe gel electrophoresis and blotting procedures in great detail (the reader is referred to references [1, 2]), a few points on these procedures which are relevant to chemiluminescent detection are given below:

1. Prepare polyacrylamide gel, preferably of 0.75-mm thickness and a strength between 6% and 12% acrylamide.
   Note: If a large size range of proteins should be blotted, a gel of about 7% acrylamide is to be preferred as proteins from 30 to about 120 kDa transfer well from this gel.

2. Prepare the samples by heating an appropriate amount of µg protein mixed with 2×loading buffer (ratio 1:1) for 2–5 min at 100 °C.
   Note 1: For tissue extracts, use about 20–40 µg/well; for a specific protein this can be much less, even into the picogram range.
   Note 2: Do not forget to load one lane with a protein molecular weight marker. These can be bought in different sizes and are even available pre-stained or biotin-labeled. The latter can be very handy when chemiluminescent detection is going to be used as the marker will be visible on the same image as the protein(s) of interest.

3. Apply samples to the gel and run (current, 10–15 mA).

4. When the run is finished (usually when the bromophenol blue dye has reached the bottom of the running gel), remove the glass plates and transfer the gel to the electroblotting apparatus.

5. Set up the transfer by placing the gel on a piece of membrane cut to the size of the gel and place two or three pieces of filter paper (cut to the size of the gel) soaked in transfer buffer on either side. Then, either wet or semi-dry blotting (refer to the manufacturer's instructions for setting up a transfer) can be used. Both can be used with equal efficiency for proteins in a range between 25–120 kDa. When large proteins (>120 kDa) need to be transferred, a semi-dry blotting apparatus is not recommended since transfer time will become too long and buffer depletion becomes a problem. When quantitative data are required, the use of PVDF membrane is recommended because of its superior binding properties (especially for smaller proteins).

   Note: When using PVDF, do not forget to pre-wet the membrane with methanol and then wash with distilled water (make sure all methanol is removed, otherwise high background may result). If the membrane is not completely gray in color after immersing in methanol, use a different piece. Make sure the PVDF membrane does not dry out at any time during the subsequent procedures. If it does, however, repeat the methanol step.

6. When the transfer is complete, check transfer by staining the membrane with Ponceau S. To do this, incubate the membrane 5 min in the Ponceau S solution.

7. Remove the Ponceau S solution and destain the blot in water for a few minutes.

8. Check whether transfer has occurred and, if necessary, mark the position of the molecular weight standards with a soft pencil.

   Note: If required, the partially destained gel can be photographed.

9. Continue to destain with water for about 15 min.

2.2.3 Antibody binding

1. Block the membrane by incubating it in 5% blocking buffer for at least 1 h at room temperature in a tray whilst shaking (use 50 ml/100 cm^2).

2. Dilute the first antibody in 1% blocking buffer (about 10 ml/100 cm^2) in the appropriate dilution (usually between 1:100 and 1:2000; see below) and incubate the blot with this solution. To minimize the amount used, we recommend placing the blot with antibody dilution in a heat-sealable

bag or in a bottle that fits a rotating oven (protein-side up) and incubate for 1 h at room temperature whilst shaking, or in the rotating oven.

Note: The first antibody dilution can be optimized by spotting an amount of the protein (extract) of interest onto pieces of membrane and going through this protocol using different first antibody dilutions (e.g., 1:100, 1:250, 1:500, 1:1000, 1:2500, 1:5000), and a fixed amount of second antibody. Following detection, the concentration which gives the best signal-to-noise ratio can be selected. Similarly, the optimal concentration of second antibody can be determined at a fixed first antibody concentration.

3. Discard the antibody dilution and wash the membrane 6×5 min in 1×TBST (50 ml/100 cm^2)
4. Dilute the second antibody (complexed with AP or HRP/POD) in 1% blocking buffer (about 10 ml/100 cm^2; usual dilution between 1:500 and 1:10000) and incubate the blot 30 min–1 h at room temperature with this solution as described for the first antibody.
5. Discard the second antibody dilution and wash the membrane 6×5 min in 1×TBST (50 ml/100 cm^2).
6. Drain the 1×TBST from the membrane and proceed to the detection protocol.

2.2.4 Detection

■ **Protocol 1 (luminol-type substrate, Fig. 1).**
Although a number of luminol-based methods are available, this protocol describes the use of the Lumi-Light^{plus} substrate, Boehringer Mannheim. The advantage of this substrate is that the signal persists for hours, which allows time to take a number of exposures (see Fig. 1).

1. Prepare the substrate solution by mixing the Lumi-Light^{plus} luminol/enhancer solution 1:1 with the Lumi-Light^{plus} peroxide solution (stable for about 24 h at room temperature).
2. Incubate the membrane 5–30 min in the substrate solution. This can be conveniently done by placing the membrane (protein-side up) on a piece of Saran Wrap and pouring the substrate solution onto the membrane. Make sure that the entire membrane is covered!

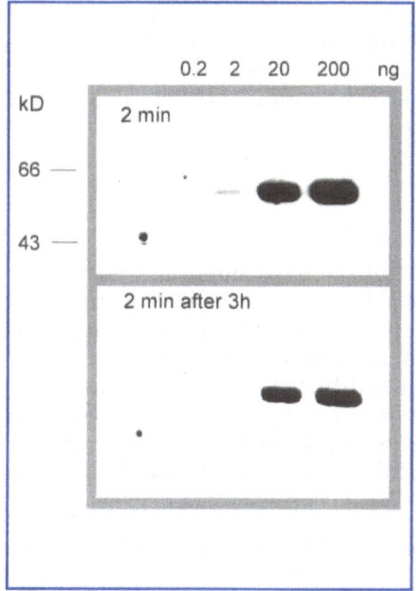

Fig. 1. Different amounts of a 55 kD 6×His tagged protein (indicated above the image) were run on a 7.5% polyacrylamide gel which was blotted onto PVDF membrane. The membrane was incubated with monoclonal anti-6×His (mouse) (1:1000, overnight at 4 °C) and thereafter with anti-mouse IgG-POD (Lumi-Light^plus Western blotting kit). To detect the bands the Lumi-Light^plus substrate solution was left on the blot for 30 min at room temperature and then a 2 min exposure was taken (*top panel*). Three hours later another 2 min exposure still showed up the same bands at a slightly lower intensity (*lower panel*; both panels have the same Display Range 208 to 3136)

3. Remove the substrate solution, drain excess, and wrap the membrane in Saran Wrap.
4. Expose to the Lumi-Imager (or film).

■ **Protocol 2 (dioxetane-type substrate)**
1. Dilute Nitro-Block reagent (Tropix) 1:20 in assay buffer and incubate the membrane for 5 min in this solution (20 ml/ 100 cm^2).
2. Dilute the substrate (CSPD® or CDP-*Star*™) 1:100 in assay buffer (5 ml is sufficient for 100 cm^2).
   Note: CDP-Star will give a more intense signal in a shorter time (so more sensitivity) but does not last as long as the CSPD signal. The latter will give similar exposures up to 4–5 days after the first application of the substrate.
3. Place the membrane on a piece of Saran Wrap (protein-side up), pour the diluted substrate on top, making sure that the membrane is completely covered. If necessary, cover with another piece of Saran Wrap (but do not press it on).

4. Leave for 5 min, drain excess substrate solution, and wrap the membrane in Saran Wrap. In the case of CSPD, preincubation for 15–30 min at 37 °C may increase light output.
5. Expose to the Lumi-Imager (or to film, example see Chapter 1, Fig. 8).

2.2.5 Quantifying the signal

Quantifying the signal is one of the most important goals after a succesful Western blot has been performed. Although this can be done on film, this is prone to error because the exposure time has to be chosen such that it is within the linear range of the film. Using a Lumi-Imager bypasses these problems as it is linear over a larger range (1:10 000) than film (1:100). The LumiAnalyst software is well suited to quantify the intensities of bands quickly. Some remarks on its use are given below:

1. As an initial exposure time, use 30 s–5 min when a luminol-type substrate has been used. If one of the dioxetane substrates has been used, longer exposure times will usually be necessary when compared to luminol-type substrates (start with 5–15 min). The result will give a clear indication of whether a shorter or longer exposure is needed.
2. Setting the baseline is an important step in quantification which should not be overlooked. "Join valleys" is usually a good starting point, whereas is some cases (e.g., when there is an uneven background) the "from image" baseline is better [3].
3. Check the extent of the band using the "Band ID" option with "Show band extents" selected [3]. Adjust if necessary using the "Lane Profile" option [3]. Failure to do this may lead to erroneous data because background will be included in the calculation.
4. If the molecular weight also needs to be determined, make sure that bands of the same length are seen by the program as such. If not, this can be corrected using the "Slant" option.

2.3 Results

Typical exposure times are between 10 s and 10 min on the Lumi-Imager when luminol-type substrates are used but, as mentioned, these
can be longer. When dioxetane substrate has been used, exposure
times are typically between 15 and 60 min, but may be longer (the
CSPD signal will last for a few days). Using the Lumi-Imager gives
more leeway in the exposure time since strong and weak signals can
be reliably quantified on the same exposure. Film quickly saturates
when compared to the Lumi-Imager image, resulting in erroneous
data. Therefore, one should always check that a film exposure which
has to be quantified is exposed within the linear range of the film.

2.4 Troubleshooting

As usual, lots of things can go wrong during the antibody
binding and subsequent detection, but the most frequently encountered problem is a high and/or uneven background. One
of the most common causes of uneven background, however, is
performing the blocking and washing steps during the binding
and detection in too little buffer and/or too small a tray (do
not use bottles in a rotating oven in these steps). The membrane should be able to move freely in the tray and be well immersed in the buffer. Another cause of high background could
be that the antibody concentration was too high, but this
should come out during the optimization protocol.

In the case of luminol-type substrates, it is possible that,
instead of a "black" band, a completely transparent ("white")
area is seen. One of the reasons for this is that there is so
much HRP/POD in this position that the substrate is completely used up in the time it takes to set up the exposure. Solutions
to this problem are to dilute the second antibody further or apply less protein to the gel.

References

1. AUSUBEL FM, BRENT R, KINGSTON RE, MOORE DD, SEICHMAN JG, SMITH JA, STRUHL K (1987) Analysis of proteins. In: Current protocols in molecular biology. Wiley, New York (Chap. 10)
2. HARLOW E, LANE D (1988) Immunoblotting. In: Antibodies: a laboratory manual. Cold Spring Harbor Laboratory, Cold Spring Harbor, New York, pp. 471–510
3. LumiAnalyst Reference guide (1997) Boehringer Mannheim, Mannheim, Version 2.0

3 Southern blotting

O. Bakker

INTRODUCTION

Southern blotting is one of the most widely used techniques for studying gene structure and mutations. Its basic principle is the detection of a DNA of interest on a solid support, mostly a nylon or nitrocellulose membrane, using a labeled probe. This probe is either DNA or RNA which can be radio-actively or non-radioactively labeled. The latter label is used more and more and this protocol aims at helping the reader in her/his efforts to use non-radioactive labels (in this case digoxigenin) for their research purpose.

3.1 Outline

A typical experiment consists of firstly purifiying the DNA from blood cells, tissues, or cell culture cells, which is then run on an agarose gel to separate the fragments. After the desired separation, the gel is "blotted" onto nylon membrane which is then prehybridized and hybridized with the probe of interest. Following hybridization, the blot is washed to remove non-specific binding and the probe is detected. In the case of digoxigenin-based chemiluminesence, which is the focus of this protocol, this detection can be done on film or using the recently developed Lumi-Imager (Boehringer Mannheim, Mannheim, Germany).

3.2 Materials

3.2.1 Equipment

- Agarose gel electrophoresis equipment (horizontal or vertical)
- If possible, blotting equipment (vacuum, electrical, or pressure)
- Rotating hybridization oven with matching bottles
- If available, a Lumi-Imager

3.2.2 Probe

- Materials to label a DNA or RNA probe with digoxigenin (e.g., those from Boehringer Mannheim; see also "Procedure").

3.2.3 Buffers

10×TBE
890 mM Tris base
890 mM Boric acid
20 mM EDTA

Loading buffer (5×)
5×TBE buffer
50% Glycerol
0.01% Bromophenol blue

20×SSC (Sodium saline citrate)	
3 M NaCl	
300 mM Sodium citrate	pH 7.0

Standard hybridization buffer + 50% formamide
5×SSC
50% Deionized formamide
0.1% Sodium lauroylsarcosine
0.02% SDS (Sodium dodecyl sulfate)
2% Blocking reagent

Buffer 1
100 mM Maleic acid   pH 7.5 (adjust with NaOH)
150 mM NaCl

Blocking reagent stock solution
Dissolve 10 g blocking reagent in 100 ml of Buffer 1 by stirring and
heating to 60 °C until dissolved (about 1 h), then autoclave

Buffer 2
Dilute blocking reagent stock solution 1:10 with Buffer 1

Wash buffer
0.3% Tween 20 in Buffer 1
Do not autoclave

Buffer 3
100 mM Tris   pH 9.5
100 mM NaCl

3.3 Procedure

3.3.1 DNA isolation

DNA can be isolated in a number of ways [1, 2]. Purification
using columns usually gives good DNA, but when very high
molecular weight DNA (about 50 kb) is desired, it is advisable
to use the "old-fashioned" method of phenol extraction [1, 2].

3.3.2 Gel electrophoresis and blotting

Although it is outside the scope of this protocol to describe gel electrophoresis and blotting procedures in great detail (the reader is referred to references [1, 2]), a few points on these procedures which are relevant to chemiluminescent detection are given below:

1. Prepare a 0.8%–1.5% agarose gel in 1×Tris-borate + EDTA (TBE). The gel strength depends on the lengths of the fragments to be separated.
   Note: For long DNAs, it may help to use a vertical agarose gel system (Owl Scientific, Woburn, MA, USA).

2. Apply 5–10 µg DNA (if genomic) or less (about 0.1 µg) when cloned fragments are used in loading buffer on the gel and run using 1×TBE buffer.

3. When the run is complete (usually when the bromophenol blue has reached the bottom of the gel), soak the gel in 0.25 M HCl for 30 min to depurinate the DNA. Failure to do so may result in bad transfer of the larger fragments (>5 kb).

4. Denature the DNA in the gel using 0.5 M NaOH/1.5 M NaCl by slowly shaking it in about 10 gel volumes for 15 min (repeat once).

5. Neutralize gel in 0.5 M Tris pH 7.0/1.5 M NaCl by slowly shaking it about 10 gel volumes for 15 min (repeat once).
   Note: The color of the bromophenol blue should change to yellow during the denaturation step 4 and become blue again during neutralization.

6. Set up the transfer. This can be done using 20×SSC or alkaline blotting and any of the available techniques [1, 2] (capillary, semi-dry, pressure, or vacuum). For the membrane, we recommend the Boehringer Mannheim nylon membrane which is function-tested for use in chemiluminescent assays. Pre-wetting the membrane with water prior to setting up the transfer is also recommended.

7. When the transfer is complete (dependent on the method), fix the DNA to the membrane using UV irradiation (timing depends on specific equipment) or baking (120 °C, 30 min).

3.3.3 Probe synthesis

1. *DNA probe:* A DNA probe can be prepared using nick translation or polymerase chain reaction (PCR) [3]. In both cases, it is possible to incorporate DIG-labeled dUTP using commercial kits or labeling mixtures.

2. *RNA probe:* These are mostly prepared using the sequences of interest cloned in a vector containing initiation sites for one of the RNA polymerases T3, T7, or SP6. In all cases, it is possible to incorporate DIG-labeled UTP which the polymerases will do very efficiently [3]. If possible, RNA labeling is the preferred labeling method as it yields a lot of highly labeled probe.

   Note: If no probe in an "RNApolymerase" vector is available, it is possible to use composite primers consisting of the initiation site of the polymerases and your gene-specific sequences (Fig. 1) to prepare a PCR fragment from which a RNA probe can be synthetized.

3. *Check labeling efficiency:* Labeling efficiency can be checked using, for instance, the DIG Quantification Test Strips of Boehringer Mannheim. Another way is by spotting 1 µl of the labeling mix onto a small piece of membrane, fixing it using UV irradiation, and going through a quick detection protocol (10 min blocking, 10 min anti-DIG-AP, two 5-min washes, and then detection). If an intense black spot appears on the film after a 1-min exposure, the probe has been labeled well and can be used.

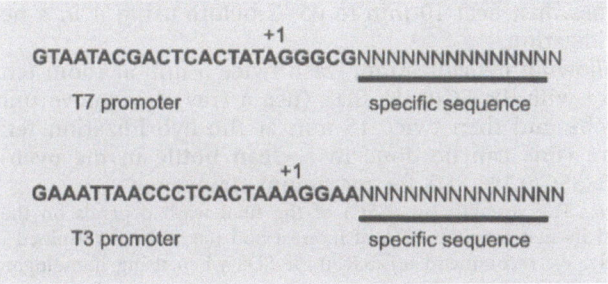

Fig. 1. Structure of the composite primers. If oligonucleotides containing the sequences of the T7 or T3 promoter coupled to the specific sequences are made, they can be used to generate templates for the synthesis of a RNA probe using any DNA fragment. It is even possible, for instance, to make the forward primer T7-dependent and the reverse T3-dependent so that both antisense and sense probes can be synthetized from one fragment

3.3.4 Hybridization

1. The hybridization mixture can in principle be any of the usual ones [1–3], but for best results the use of either Standard Hybridization buffer + 50% formamide [3] or DIG-Easy Hyb [3] (a urea-based mix; Boehringer Mannheim) is recommended.

2. Prehybridize the blot in hybridization mixture without probe for at least 1 h at the desired hybridization temperature in a bottle in a rotating hybridization oven. Other setups can also be used, but a rotating oven gives best results (not only for non-radioactive methods).
   Note: If the membrane is dry prior to starting prehybridization, first wet it with a little sterile distilled water.

3. When a double-stranded DNA probe is going to be used, denature this by heating for 5–10 min at >95°C and place immediately on ice. Dilute the probe at a concentration of 25 ng/ml (DNA probe) or 50–100 ng/ml (RNA probe) in the prewarmed hybridization mix and hybridize for the desired time (usually overnight) at the desired temperature (37–50°C [DNA probe] or 65°C [RNA probe]).
   Note: The best probe concentration and hybridization temperatures need to be determined empirically for every new probe. Furthermore, checking background using a mock hybridization is recommended [3] (too much probe will give bad background).

4. When the hybridization is complete, save the hybridization solution and freeze it at –20°C. It can be reused up to four times. Just heat 10 min to 65°C before using it in a new hybridization.

5. Following hybridization, wash twice 5 min at room temperature with 2×SSC/0.1% SDS (use a tray) to remove unbound probe and then twice 15 min at the hybridization temperature (this can be done in a clean bottle in the oven) with 0.5×SSC/0.1% SDS for maximum stringency.
   Note: The strength (in ×SSC) of the final wash depends on the probe and its match to the RNA of interest and has to be determined empirically. We recommend 0.1×SSC/0.1% SDS when using homologous RNA probes.

6. After washing, soak the membrane in wash buffer for 5 min before proceeding to detection.

3.3.5 Detection

1. Incubate the membrane in Buffer 2 (100 ml/100 cm^2) for at least 30 min (longer is not a problem) in a tray whilst shaking. The use of a rotating oven in this step is not recommended.
2. Dilute anti-DIG-alkaline phosphatase (anti-DIG-AP) 1:10 000 in Buffer 2.
3. Incubate the membrane (DNA-side up) 30 min with the anti-DIG-AP dilution. This can be done in a rotating oven, in which case about 5–7 ml DIG-AP dilution is sufficient.
4. Wash the membrane twice (15 min each) in a tray with wash buffer (100 ml/100 cm^2).
5. Rinse the membrane in ample Buffer 3 in order to remove the last traces of detergent.
6. Incubate the membrane in fresh Buffer 3 for 5 min.
7. Dilute the substrate (CSPD or CDP-Star) 1:100 in Buffer 3 (5 ml is sufficient for an average size gel).
   Note: CDP-Star will give a more intense signal in a shorter time (so more sensitivity), but does not last as long as the CSPD signal. The latter will give similar exposures up to 4–5 days after the first application of the substrate.
8. Place the membrane on a piece of Saran Wrap (DNA-side up), pour the diluted substrate on top, making sure that the membrane is completely covered. If necessary, cover with another piece of Saran Wrap (but do not press it on).
9. Leave for 5 min, then expose to the Lumi-Imager (or to film).

3.3.6 Quantifying the signal

Quantifying the signal can be one of the goals after a succesful Southern blot has been performed, for instance, in the case of restriction site polymorphisms. Although this can be done on film, this is prone to error because the exposure time has to be chosen such that it is within the linear range of the gel. Using a Lumi-Imager bypasses these problems as it is linear over a larger range (1:10 000) than film (1:100). The LumiAnalyst software is well suited to quantify the intensities of bands quickly. Some remarks on its use are given below:

1. As an initial exposure time, use 5–15 min when cloned fragments are used or 15–30 min when looking at genomic DNA. The result will give a clear indication of whether a shorter or longer exposure is needed.
2. Setting the baseline is an important step in the quantification which should not be overlooked. "Join valleys" is usually a good starting point, whereas is some cases (e.g., when there is an uneven background) the "from image" baseline is better [3].
3. Check the extent of the band using the "Band ID" option with "Show band extents" selected [3]. Adjust if necessary using the "Lane Profile" option [3]. Failure to do this may lead to erroneous data since background will be included in the calculation.

3.4 Results

Typical exposure times are between 15–120 min on the Lumi-Imager when using genomic DNA (Fig. 2) or 5–15 min when looking at cloned DNA. Using the Lumi-Imager gives more leeway in the exposure time as strong and weak signals can be reliably quantified on the same exposure.

3.5 Troubleshooting

As usual, lots of things can go wrong during a hybridization and the subsequent detection, but the most frequently encountered problem is a high and/or uneven background. One of the most common causes of uneven background, however, is performing the blocking and washing steps during the detection in too little buffer and/or too small a tray (*remember:* do not use bottles in a rotating oven in these steps). The membrane should be able to move freely in the tray and be well immersed in the buffer. Another cause of especially high background could be that the probe concentration was too high, but this problem can be overcome by performing a mock hybridization

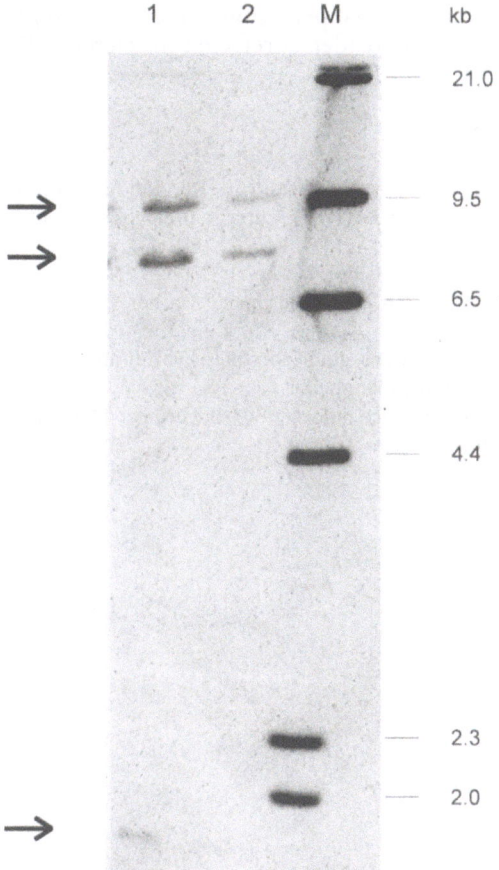

Fig. 2. Detection of the LDL receptor gene in human genomic DNA. 5 μg of EcoRI-digested human genomic DNA (two different samples) were run on a 0.8% agarose gel and blotted using alkaline blotting onto Boehringer Mannheim nylon membrane. The marker (lane *M*) is a DIG-labeled marker II of Boehringer Mannheim. Hybridization was at 55°C overnight with a LDL receptor RNA probe (at 50 ng/ml) spanning the first 11 exons. Final wash stringency was 0.1×SSC/0.1% SDS and the exposure time was 15 min using CDP-Star. *Arrows* indicate the bands which were expected from the restriction map

before the real one is performed. For an extensive trouble-shooting guide, the reader is referred to reference [3].

References

1. Ausubel FM, Brent R, Kingston RE, Moore DD, Seichman JG, Smith JA, Struhl K (1987) Current protocols in molecular biology. Wiley, New York
2. Sambrook J, Fritsch EM, Maniatis T (1989) Molecular cloning: a laboratory manual, 2nd edn. Cold Spring Harbor Laboratory, Cold Spring Harbor, New York
3. The DIG system user's guide for filter hybridisation (1995) Boehringer Mannheim GmbH, Mannheim
4. LumiAnalyst Reference guide (1997) Boehringer Mannheim, Mannheim, Version 2.0

4 Northern blotting

O. Bakker

INTRODUCTION

Northern blotting is one of the most widely used techniques for studying gene expression. Its basic principle is the detection of a RNA of interest on a solid support, mostly a nylon or nitrocellulose membrane, using a labeled probe. This probe is either DNA or RNA which can be radioactively or non-radioactively labeled. The latter label is used more and more and this protocol aims at helping the reader in her/his efforts to use non-radioactive labels (in this case digoxigenin) for research purposes.

4.1 Outline

A typical experiment consists of firstly purifiying the RNA from tissues or cells, which is then run on an agarose gel to separate the fragments. After the desired separation, the gel is "blotted" onto nylon membrane which is then prehybridized and hybridized with the probe of interest. Following hybridization, the blot is washed to remove non-specific binding and the probe is detected. In the case of digoxigenin-based chemiluminesence, which is the focus of this protocol, this detection can be done on film or using the recently developed Lumi-Imager.

4.2 Materials

4.2.1 Equipment

- Agarose gel electrophoresis equipment (horizontal or vertical).
- If possible, blotting equipment (vacuum, electrical, or pressure).
- Rotating hybridization oven with matching bottles.
- If available, a Lumi-Imager.

4.2.2 Probe

- Materials to label the DNA or RNA probe with digoxigenin (e.g., those from Boehringer Mannheim; see also procedure).

4.2.3 Buffers

Important: To prevent RNase contamination, always use diethylpyrocarbonate (DEPC)-treated water for the preparation of buffers. All buffers should be autoclaved before use unless otherwise indicated.

10×MOPS
200 mM Morpholinopropane-   pH 7.0 with sodium hydroxide
sulfonic acid
50 mM Sodium acetate
10 mM EDTA

Loading buffer (*make fresh*)
250 µl Deionized formamide
83 µl Formaldehyde 37%
50 µl 10×MOPS buffer
50 µl Glycerol
0.01% Bromophenol blue
Fill up to 500 µl with water

20×SSC (Sodium saline citrate)
3 M NaCl
300 mM Sodium citrate   pH 7.0

Standard hybridization buffer + 50% formamide
5×SSC
50% Deionized formamide
0.1% Sodium lauroylsarcosine
0.02% SDS
2% Blocking reagent

Buffer 1
100 mM Maleic acid   pH 7.5 (adjust with NaOH)
150 mM NaCl

Blocking reagent stock solution
Dissolve 10 g blocking reagent in 100 ml of Buffer 1 by stirring and
heating to 60 °C until dissolved (about 1 h), then autoclave

Buffer 2
Dilute blocking reagent stock solution 1:10 with Buffer 1

Wash buffer
0.3% Tween 20 in Buffer 1
Do not autoclave

Buffer 3
100 mM Tris   pH 9.5
100 mM NaCl

4.3 Procedure

4.3.1 RNA isolation

RNA can be isolated in a number of ways. To purifiy total RNA single-step solutions (e.g., TriPure, Boehringer Mannheim) are available. Very often, using polyA$^+$ RNA is better, especially with mRNAs which are of similar size to the ribosomal RNAs (about 2 and 6 kb). PolyA$^+$ RNA can be directly, and conveniently, purified from (small amounts of) tissues or cells using methods based on magnetic particles (e.g., mRNA isolation kit, Boehringer Mannheim; PolyATract, Promega Corporation).

4.3.2 Gel electrophoresis and blotting

Although it is outside the scope of this protocol to describe gel electrophoresis and blotting procedures in great detail (the reader is referred to references [1, 2]), a few points on these procedures which are relevant to chemiluminescent detection are given below:

1. Prepare a 1% agarose gel containing 2.2 M formaldehyde (in a fume hood).
   Note: For long mRNAs it may help to use a vertical agarose gel system (Owl Scientific, Woburn, MA, USA). In this case, use 0.7 M formaldehyde to prevent the wells from breaking.
2. Prepare the samples by heating 10 µg total RNA or 0.5–1 µg polyA$^+$ RNA in loading buffer.
3. Apply samples to the gel and run using 1×MOPS buffer.
4. When the run is finished (usually when the bromophenol blue has reached the bottom of the gel), soak the gel in two changes of sterile 20×SSC for 15–20 min to remove the formaldehyde. Failure to do so may result in bad transfer.
5. Set up the transfer. This can be done using any of the available techniques [1, 2] (capillary, semi-dry, pressure, or vacuum). For the membrane, we recommend the Boehringer Mannheim nylon membrane which is function-tested for use in the chemiluminescent assays. Good results have also been obtained with Nytran NY13 (Schleicher and Schuell, Dassel,

Germany) [5–8]. Pre-wetting the membrane with water prior setting up the transfer is recommended.

6. When the transfer is complete (depending on the method), fix the RNA to the membrane using UV irradiation (timing depends on specific equipment) or baking (120 °C, 30 min).

4.3.3 Probe synthesis

1. *DNA probe:* A DNA probe can be prepared using nick translation or polymerase chain reaction (PCR) [3]. In both cases, it is possible to incorporate DIG-labeled dUTP using commercial kits or labeling mixtures.

2. *RNA probe:* These are mostly prepared using the sequences of interest cloned in a vector containing initiation sites for one of the RNA polymerases T3, T7, or SP6. In all cases, it is possible to incorporate DIG-labeled UTP, which the polymerases will do very efficiently [3]. If possible, RNA labeling is the preferred labeling method as it yields at lot of highly labeled probe.

   Note: If no probe in an "RNApolymerase" vector is available, it is possible to use composite primers consisting of the initiation site of the polymerases and your gene-specific sequences (Fig. 1) to prepare a PCR fragment from which an RNA probe can be synthetized.

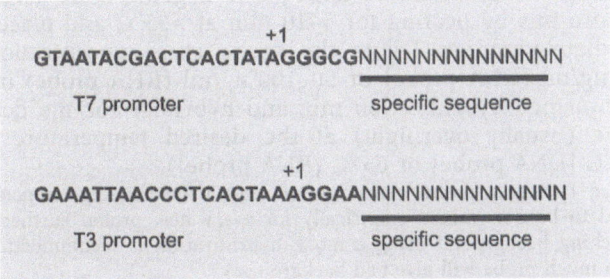

Fig. 1. Structure of the composite primers. If oligonucleotides containing the sequences of the T7 or T3 promoter coupled to the specific sequences are made, they can be used to generate templates for the synthesis of a RNA probe using any DNA fragment. It is even possible, for instance, to make the forward primer T7-dependent and the reverse T3-dependent, so that both antisense and sense probes can be synthetized from one fragment

3. *Check labeling efficiency:* Labeling efficiency can be checked using, for instance, the DIG Quantification Test Strips of Boehringer Mannheim. Another way is by spotting 1 µl of the labeling mix onto a small piece of membrane, fixing it with UV irradiation, and going through a quick detection protocol (10 min blocking, 10 min anti-DIG-AP, two 5-min washes, and then detection). If an intense black spot appears on film after a 1-min exposure, the probe has been labeled well and can be used.

4.3.4 Hybridization

1. The hybridization mixture can in principle be any of the usual ones [1–3], but for best results the use of either Standard Hybridization buffer + 50% formamide [3] or DIG-Easy Hyb [3] (an urea-based mix; Boehringer Mannheim) is recommended.
2. Prehybridize the blot in hybridization mixture without probe for at least 1 h at the desired hybridization temperature in a bottle in a rotating hybridization oven. Other setups can also be used but a rotating oven gives best results (not only for non-radioactive methods).
   Note: When the membrane is dry, prior to starting prehybridization, first wet it with a little sterile distilled water.
3. When a double-stranded DNA probe is going to be used, denature this by heating for 5–10 min at >95°C and place immediately on ice. Dilute the probe at a concentration of 25 ng/ml (DNA probe) or 50–100 ng/ml (RNA probe) in the prewarmed hybridization mix and hybridize for the desired time (usually overnight) at the desired temperature (37–50°C [DNA probe] or 65°C [RNA probe]).
   Note: The best probe concentration and hybridization temperatures need to be determined empirically for every new probe. Furthermore, checking background using a mock hybridization is recommended [3] (too much probe will give bad background).
4. When the hybridization is finished, save the hybridization solution and freeze it at −20°C. This can be reused up to four times. Just heat 10 min to 65°C before using it in a new hybridization.

5. Following hybridization, wash twice for 5 min at room temperature with 2×SSC/0.1% SDS (use a tray) to remove unbound probe and then twice 15 min at the hybridization temperature (this can be done in a clean bottle in the oven) with 0.1×SSC/0.1% SDS for maximum stringency.

   Note: The strength (in ×SSC) of the final wash depends on the probe and its match to the RNA of interest and has to be determined empirically. We recommend 0.1×SSC/0.1% SDS when using homologous RNA probes.

6. After washing, soak the membrane in wash buffer for 5 min before proceeding to detection.

4.3.5 Detection

1. Incubate the membrane in Buffer 2 (100 ml/100 cm^2) for 30–60 min in a tray whilst shaking (*remember:* work RNase free [1, 2]). The use of a rotating oven in this step is not recommended.

2. Dilute anti-DIG-alkaline phosphatase (anti-DIG-AP) 1:10 000 in Buffer 2.

3. Incubate the membrane (RNA-side up) 30 min with the anti-DIG-AP dilution. This can be done in a rotating oven, in which case about 5–7 ml DIG-AP dilution is sufficient.

4. Wash the membrane twice (15 min each) in a tray with wash buffer (100 ml/100 cm^2).

5. Rinse the membrane in ample Buffer 3 in order to remove the last traces of detergent.

6. Incubate the membrane in fresh Buffer 3 for 5 min.

7. Dilute the substrate (CSPD or CDP-Star) 1:100 in Buffer 3 (5 ml is enough for an average size gel).

   Note: CDP-Star will give a more intense signal in a shorter time (so more sensitivity) but does not last as long as the CSPD signal. The latter will give similar exposures up to 4–5 days after the first application of the substrate.

8. Place the membrane on a piece of Saran Wrap (RNA-side up), pour the diluted substrate on top, making sure that the membrane is completely covered. If necessary, cover with another piece of Saran Wrap (but do not press it on).

9. Leave for 5 min, then expose to the Lumi-Imager (or to film).

4.3.6 Quantifying the signal

Quantifying the signal is one of the most important goals after a succesful Northern blot has been performed. Although this can be done on film, this is prone to error because the exposure time has to be chosen such that it is within the linear range of the gel. Using a Lumi-Imager bypasses these problems as it is linear over a larger range (1–10 000) than film (1–100). The LumiAnalyst (Boehringer Mannheim, Mannheim, Germany) software is well suited to quantify the intensities of bands quickly. Some remarks on its use it are given below:

1. As an initial exposure time, use 5–15 min. The result will give a clear indication of whether a shorter or longer exposure is needed.
2. Setting the baseline is an important step in the quantification which should not be overlooked. "Join valleys" is usually a good starting point, whereas in some cases (e.g., when there is an uneven background) the "from image" baseline is better.
3. Check the extent of the band using the "Band ID" option with "Show band extents" selected. Adjust if necessary using the "Lane Profile" option. Failure to do this may lead to erroneous data since background will be included in the calculation.

4.4 Results

Typical exposure times are between 5–60 min on the Lumi-Imager-when using total RNA, or from 5 s when using polyA$^+$ RNA and looking at an abundant message (Fig. 2a). Using the Lumi-Imager gives more leeway in the exposure time since strong and weak signals can be reliably quantified on the same exposure. Film quickly saturates when compared to the Lumi-Imager image, resulting in erroneous data (see Chapter 1, Fig. 4).

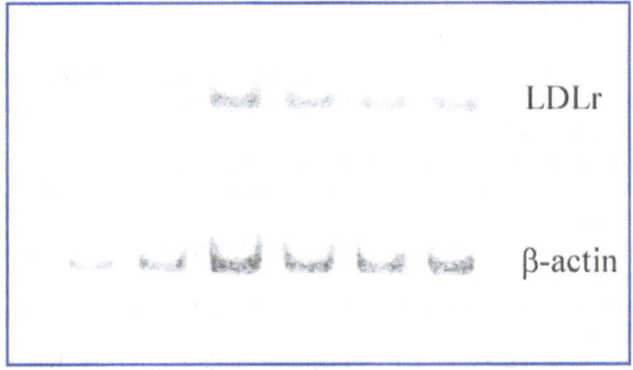

Fig. 2. Northern blot of rat liver messenger RNA. PolyA$^+$ RNA was purified from rat livers and 3 μg was applied per lane. The gel was run on a vertical agarose gel system (Owl Scientific) and blotted onto Boehringer Mannheim nylon membrane using pressure blotting (PosiBlot, Stratagene). The blot was prehybridized and hybridized at 65 °C in DIG Easy Hyb using RNA probes for both the LDL receptor (LDLr) and β-actin. Final stringency was 0.1×SSC/0.1% SDS at 65 °C. Detection was with CDP-Star. Exposure times were 45 sec for LDLr and 5 sec for β-actin

4.5 Troubleshooting

As usual, lots of things can go wrong during a hybridization and the subsequent detection, but the most frequently encountered problem is a high and/or uneven background. One of the most common causes of uneven background, however, is performing the blocking and washing steps during the detection in too little buffer and/or too small a tray (*remember:* do not use bottles in a rotating oven in these steps). The membrane should be able to move freely in the tray and be well immersed in the buffer. Another cause of especially high background could be that the probe concentration was too high, but this problem can be overcome by performing a mock hybridization before the real one is performed. For an extensive troubleshooting guide, the reader is referred to reference [3].

References

1. Ausubel FM, Brent R, Kingston RE, Moore DD, Seichman JG, Smith JA, Struhl K (1987) Current protocols in molecular biology. Wiley, New York
2. Sambrook J, Fritsch EM, Maniatis T (1989) Molecular cloning: a laboratory manual, 2nd edn. Cold Spring Harbor Laboratory, Cold Spring Harbor, New York
3. The DIG system user's guide for filter hybridization (1995) Boehringer Mannheim, Mannheim
4. LumiAnalyst reference guide (1997) Boehringer Mannheim, Mannheim, vol. 2.0
5. Boelen A, Platvoet-ter Schiphorst MC, Bakker O, Wiersinga WM (1995) J Endocrinol 146:475–483
6. Boelen A, Maas MAW, Lowik CWGM, Platvoet-ter Schiphorst MC, Wiersinga WM (1996) Endocrinology 137:5250–5254
7. van der Wal AMG, Bakker O, Wiersinga WM (1998) Int J Biochem 30:209–215
8. Hudig F, Bakker, Wiersinga WM (1998) Metabolism, in press

Suppliers

- Amersham International plc, Amersham place, Little Chalfont, Buckingham HP7 9NA, UK
- Boehringer Mannheim GmbH, Sandhofer Strasse 116, D-68305 Mannheim, Germany
- DAKO A/S, Produktionsvej 42, 2600 Glostrop, Denmark
- Owl Scientific Inc., 10 Commerce Way, Woburn, MA 01801, USA
- Promega Corporation, 2800 Woods Hollow Rd, Madison, WI 53711–5399, USA
- Schleicher & Schuell, Postfach 4, D-37582 Dassel, Germany
- Tropix, 47 Wiggins Avenue, Bedford, MA 01730, USA

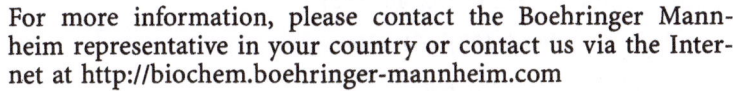

For more information, please contact the Boehringer Mannheim representative in your country or contact us via the Internet at http://biochem.boehringer-mannheim.com

- Argentina   Tel.: 541 954 5555

- Australia   Tel.: (02) 9899 7999

- Austria   Tel.: (01) 277 87

- Belgium   Tel.: (02) 247 4930

- Brazil   Tel.: +55 (11) 3666 3565

- Canada   Tel.: (450) 686 7050; (800) 361 2070

- Chile   Tel.: 00 56 (2) 22 33 737 (central)
           00 56 (2) 22 32 099 (Exec)

- China   Tel.: 86 21 6427 5586

- Columbia   Tel.: 0057-1-3412797

- Czech Republic   Tel.: (0324) 45 54, 58 71-2

- Denmark   Tel.: 45 16 09 50
- Egypt   Tel.: 202 453 1133
- Finland   Tel.: (09) 429 2342
- France   Tel.: 04 76 76 30 87
- Germany   Tel.: (0621) 759 8540
- Hong Kong   Tel.: (852) 2485 7596
- India   Tel.: (22) 431 2312
- Indonesia   Tel.: 62 (021) 252 3820 ext. 755
- Iran   Tel.: 00 98 212 08 2266; 00 98 218 78 5656
- Israel   Tel.: 972 3 6 49 31 11
- Italy   Tel.: 039 247 4109-4181
- Japan   Tel.: 03 5443 5284
- Kenya   Tel.: 00254-2-74 46 77
- Kuwait   Tel.: 00965-483 26 00
- Luxembourg   Tel.: 00352-4824821
- Malaysia   Tel.: 60 (03) 261 1100
- Mexico   Tel.: (5) 227 8967
- Netherlands   Tel.: (036) 539 4911
- New Zealand   Tel.: (09) 276 4157
- Nigeria   Tel.: 00234-1-96 09 84
- Norway   Tel.: 22 07 65 00
- Philippines   Tel.: (632) 810 7246
- Poland   Tel.: +48 (22) 667 9168
- Portugal   Tel.: (01) 4171717
- Republic of Ireland   Tel.: 1 800 40 90 41
- Russia   Tel.: (49) 621 759 8636; Fax: (49) 621 759 8611
- Saudia Arabia   Tel.: +966 1 4010333

- Singapore   Tel.: 65 272 9200
- South Africa   Tel.: (011) 886 2400
- South Eastern Europe   Tel.: (01) 277 87
- South Korea   Tel.: 02 569 6902
- Spain   Tel.: (93) 201 4411
- Sweden   Tel.: (08) 404 8800
- Switzerland   Tel.: +41 (41) 799 6161
- Taiwan   Tel.: (02) 2736 7125
- Thailand   Tel.: 66 (2) 937 04 44
- United Kingdom   Tel.: (0800) 521578
- USA   Tel.: (800) 428 5433

Springer
and the
environment

At Springer we firmly believe that an international science publisher has a special obligation to the environment, and our corporate policies consistently reflect this conviction.

We also expect our business partners – paper mills, printers, packaging manufacturers, etc. – to commit themselves to using materials and production processes that do not harm the environment. The paper in this book is made from low- or no-chlorine pulp and is acid free, in conformance with international standards for paper permanency.

Springer